BRAINS

JOHNS HOPKINS
UNIVERSITY PRESS

AARHUS UNIVERSITY PRESS

brains

LEIF ØSTERGAARD

ELECTRIN

BRAINS

© Leif Østergaard
and Johns Hopkins University Press 2023
Layout and cover: Camilla Jørgensen, Trefold
Cover photograph: Poul Ib Henriksen
Publishing editor: Karina Bell Ottosen
Translated from the Danish by Heidi Flegal
Printed by Narayana Press, Denmark
Printed in Denmark 2023

ISBN 978-1-4214-4604-2 (pbk)
ISBN 978-1-4214-4605-9 (ebook)

Library of Congress Control Number: 2022938147

*Special discounts are available for bulk purchases of this
book. For more information, please contact Special Sales at
specialsales@jh.edu.*

Published in the United States by:

Johns Hopkins University Press
2715 North Charles Street
Baltimore, MD 21218-4363
www.press.jhu.edu

Published with the generous support of the
Aarhus University Research Foundation

Purchase in Denmark: ISBN 978-87-7219-020-4

Aarhus University Press
Finlandsgade 29
8200 Aarhus N
Denmark
www.aarhusuniversitypress.dk

PEER
REVIEWED

MIX
Paper
FSC FSC® C010651

CONTENTS

THE MIND OF A GENIUS

PICKLED BRAIN, ANYONE?

Cogito ergo sum. "I think, therefore I am." This quote from the French philosopher René Descartes is a perfect way to open any book about the brain. Our brains enable us to think and to remember. To feel a fragrant breeze caress our cheek. Our brains make us want to investigate the world. Perhaps your brain made you pick up this book? Whether you finish it or not, you can thank your brain for deciphering the black squiggles on this page and, hopefully, finding meaning in the words. And not to forget: Brains enable us to produce thoughts that can change the world forever.

For thousands of years, marvellous ideas and exceptional works of art have been ascribed to 'divine inspiration', but how do they arise in the brain of a genius? We do not know. When Albert Einstein died in 1955, an American pathologist named Thomas Harvey removed the eminent physicist's brain before his body was cremated. Harvey, believing Einstein's brain could explain his phenomenal ingenuity, photographed the organ, cut it into 240 pieces and made wafer-thin slices of brain tissue to study the cells under his microscope. He was sorely disappointed.

The brain of an adult male *Homo sapiens* typically weighs 1350 grams (about three pounds). To Harvey's surprise, Einstein's brain weighed just 1230 grams, so for brains (as for various other organs) size is not decisive for success. This is also good news for females of our species, whose brains are, on average, 10% smaller than male brains – like Einstein's was.

Harvey failed to find the key to Einstein's genius under the microscope, and the tissue that once hosted a brilliant mind was gradually forgotten in a cardboard box in Harvey's office. Fortunately, the large pickling jar holding Einstein's brain tissue resurfaced, and today select slices are on display at museums around the world.

Researchers are still studying Einstein's brain, but we now know that the links between brain structure, thoughts and behaviours are far more complex than Harvey could ever have imagined. The quest goes on as scientists search for the genesis of excellent ideas like theories of relativity, Facebook and pre-sliced bread.

SLICED PORRIDGE

Harvey's idea that the secret of Einstein's genius could be found in the microscopic structures in his brain had its roots in the late 1800s. An Italian physician named Camillo Golgi believed mental illness could result from brain damage, so he began to study slices of brain tissue from deceased patients to prove his theory.

But Golgi had a problem. Under a microscope, slices of brain basically look like flattened blobs of cold oat porridge, so he had to develop a method to highlight the tissue structures that might expose the mysterious workings of the mind.

In 1873, many experiments later, Golgi developed a chemical reaction that could dye or 'stain' the constituent parts of brain tissue in different shades. His process was remarkable – almost like developing an image in a darkroom and watching it materialise on photographic paper. Suddenly, Golgi's microscope revealed an extensive network of branching fibres that had characteristic patterns in different parts of the brain.

Golgi drew everything he saw under his microscope, creating veritable works of art. News of his staining technique and of the brain's complex structures quickly spread. Like many of his contemporaries, Golgi believed the nerve fibres were parts of a single, vast, unified cell.

One exception was the Spanish physician Santiago Ramón y Cajal. After seeing Golgi's drawings in 1887 and applying his technique, Ramón y Cajal realised the brain actually consists of billions of tiny individual cells. The treelike structures Golgi had described were extensions of the nerve cells we now call *neurons*.

Ramón y Cajal was also the first to discover the neuron's many *synapses* – tiny, toadstool-shaped projections on the neuron extensions that enable neurons to communicate by releasing chemical messenger molecules – *neurotransmitters*. A neuron looks a bit like a tree. Picture its lower trunk and root end – the *axon* – acting as a conduit that sends signals down into a network of extensions tipped in synapses, which are in contact with other neurons. The thick upper trunk of this tree, containing the neuron's cell nucleus, spreads into a crown of delicate branches – *dendrites* – which receive signals from the synapses of other neurons.

In 1894, Ramón y Cajal proposed that the brain stores information in the networks neurons create through their synapses, and not, as one might think, by generating new neurons for new information. His description of neurons heralded the birth of modern neuroscience.

300,000,000,000,000 CONNECTIONS

Today, we know a human brain contains about 85 billion neurons. Each of these communicates, on average, with 7,000 other neurons, which each in turn communicate with another 7,000 neurons, and so on and so forth.

By comparison, estimates say that each of Earth's roughly 7.5 billion inhabitants has a social network of 300–700 people. Put simply, it would be far more complex to understand how a brain's network operates than to map out the communication between everyone alive on the planet today – a huge challenge even for the combined resources of all the intelligence and security agencies in the world.

Now let's take a look at your brain. You have probably seen a 'pudding-bowl haircut', the kind that looks as if you simply put a pot on someone's head and cut away everything below the edge. Do *not* try this at home, but imagine that instead of scissors we took a saw and cut through the skin and the 5–10 mm of bone in your cranium. Removing the domed lid, we could bare your greyish *cerebrum*, which takes up most of the space in your skull cavity. It is covered by protective membranes, which we could gradually remove, revealing the blood vessels on its uneven surface.

The cerebrum consists of two nearly symmetrical halves, each roughly the shape and size of half a cantaloupe melon sliced lengthwise. The two halves are covered by the grey-toned cerebral *cortex*, whose *grey matter* contains the neurons' cellular nuclei and countless dendrites. The cortex is 1–4 mm thick and folds in upon itself, creating a convoluted landscape of half-inch-wide *gyri*, each ridge or *gyrus* separated from the others by deep grooves or *sulci*.

Beneath the cerebral cortex, the long neuronal axons form the brain's *white matter*, which is a profusion of biological wires enabling the various parts of the brain to communicate. These wires also carry your brain's commands to your muscles and internal organs, and they receive sensory input from your skin, eyes, ears, nose and mouth.

The wires are white because of *myelin*, a fatty substance that insulates the axons. Thanks to myelin, your neurons can send electrical impulses racing across long distances at over 200 kilometres (125 miles) per hour.

Besides its 85 billion or so neurons, your grey matter contains an equal number of star-shaped cells called *astrocytes*. Their myriad extensions connect with the brain's blood vessels and neuronal synapses. Certain parts of

Einstein's brain seem to be unusually rich in astrocytes, and scientists are still trying to understand what these enigmatic cells do – and whether they help us get bright ideas.

NEURAL PATHWAYS SWITCHING SIDES

Each half of your cerebrum is divided into *lobes*. Starting above each eye socket, the *frontal lobe* stretches from your forehead back to the *central sulcus*, a deep fissure that separates the frontal lobe from the *parietal lobe*. The central sulcus begins at the apex of the skull, slanting downwards to a point about midway between the outer corner of your eye and your outer ear canal.

Neurosurgeons quickly learn to recognise the central sulcus. They are acutely aware that damaging the gyrus in front of the central sulcus – that is, the gyrus at the rear of the frontal lobe – can cause permanent paralysis in a patient. Brain researchers call this area the *motor strip*, because you use the cortex here to perform movements, such as turning the pages of this book.

A blood clot or tumour that damages the motor strip on one side of the brain will paralyse parts of the opposite side of the body. This is because the neural pathways switch sides on their way from your cerebral cortex to your muscles. The cortex on the parietal lobe just behind the central sulcus is called the *sensory strip*. A kiss on the cheek or a stone in your shoe puts this area to work, and once again the sensory apparatus on the left half of your body sends signals to the right half of your brain, and vice versa.

Behind your eye sockets, beneath your frontal lobes,

are your *temporal lobes*. They rest on the floor of your skull, stretching back just beyond your ears. The two central sulci, which separate your frontal and parietal lobes, stop at the upper edge of your temporal lobes, about two fingers above your ears. The upper part of your temporal lobes houses the *auditory cortex*. This is where signals from your auditory nerves are analysed, letting you hear wailing babies, music, conversations and screeching brakes in traffic.

At the very back of the cerebrum, behind the parietal and temporal lobes, are your two *occipital lobes*. They treat the sensory input your retinas send back into the cerebral cortex, along your visual nerves. In other words, the *visual cortex* at the back of your head is working hard as you read the letters on this page.

THE FAMOUS FRONTAL LOBE OF PHINEAS GAGE

In 1848, a blasting foreman named Phineas P. Gage accidentally gave science valuable insight into the functions of our frontal lobes. During the great railway expansion in the north-eastern United States, Gage's crew blasted passageways through hilly areas to make way for the tracks. His crew would drill deep into the rock, then fill the holes with gunpowder and fuse lines. Gage would tamp down sand and gravel over the explosives to enhance the effect, using a tamping iron he had commissioned from a blacksmith: a solid rod four feet long and one and a half inches thick (110 cm x 3.2 cm). One end fit snugly into the drilled holes while the other was tapered so Gage could grip the fifteen-pound (6 kg) iron rod firmly with both hands.

Late one afternoon, on 13 September 1848, Gage and his crew were working south of Cavendish, Vermont, when misfortune struck. As Gage was tamping a hole, a spark flew as iron met rock. The charge exploded, transforming the tamping iron into a high-powered projectile that hit Gage just below his left cheekbone. In a split-second the long, pointed rod passed behind his left eye, up through the floor of his cranium, through his left frontal lobe, and out the top of his skull. It landed some 80 feet (25 m) away, streaked with blood and brain tissue.

Despite this horrific head trauma Gage quickly regained consciousness, and thanks to John Harlow, a resolute and accomplished local doctor, he also survived his severe burns and the festering abscesses that formed in his brain over the ensuing weeks. Gage insisted on going back to work, but although his blasting skills were intact and his speech and motor skills only slightly impaired, his *personality* was utterly transformed.

The calm and pensive Gage had lost all sense of propriety and become so crude and foul-mouthed that, as friends reportedly remarked, he was "no longer Gage". The once popular foreman no longer had the decision-making and planning skills to do his job, so the railway company was soon forced to fire him.

Gage, with his intact right frontal lobe, is seen today as an example of how other areas of the brain can compensate for cerebral-cortex damage. Over time he actually regained some of his former composure, and after several years he was able to work as a stagecoach driver and labourer. Twelve years after the accident he was suffering from frequent epileptic seizures, probably due to his brain injury, and he died on 21 May 1860 in San Francisco. In 1866, Gage's family bequeathed his skull and tamping iron to Harlow, and today both objects are on display at a museum in Boston.

WORK THAT CEREBELLUM

You have just met your cerebrum, which has a companion: the *cerebellum* or 'small brain'. Cradled in the lower, hindmost cavity of the cranium, the cerebellum is only one tenth the size of the cerebrum. One thing it does for us is regulate the commands the cerebrum sends to our muscles when we move, based on the signals that tendons and muscles return to the brain through our nerve fibres. A doctor who suspects a patient's cerebellum is malfunctioning often has the patient perform this little test:

Stretch one arm out in front of you, then flex it and touch your index finger to the tip of your nose. Easy, right? That is your cerebellum making sure the movement is even and precise. Now do the same thing with your eyes closed. You probably did a decent job, since most often the cerebellum only needs the signals from your muscles and tendons to figure out where your index fingertip or some other body part is situated.

Once you plan and initiate a complex movement, your

cerebellum helps adjust the nerve impulses going out to the relevant muscles so things go according to plan. Say you need to carry a heavy log through a forest. Your brain will not know in advance how soft the forest floor is, how heavy the log is, or how the weight will be distributed on your arms, back and legs as you carry it. Your cerebellum and cerebrum work together to fine-tune your movements.

The cerebellum is also busy computing and adjusting in teenagers going through a growth spurt. A young brain may know many movements, but during this awkward recalibration phase it takes a big effort from ground control to make a gangly adolescent body do what used to come naturally.

THE REAL RAIN MAN

A remarkable American boy named Kim Peek did not learn to walk until he was four years old. Even as an adult he could not tie his shoes or button his own shirt. His impaired motor skills were probably caused by damage to his cerebellum. Peek's cerebrum, on the other hand, was unusually large. Endowed with an exceptional memory, he inspired a character called Raymond 'Rain Man' Babbitt, played by Dustin Hoffman in the film of that name.

Peek was able to read thick volumes in about an hour, two pages at a time – the left pages with his left eye, and the right pages with his right eye. Afterwards he could recall more than 98% of the content and repeat it verbatim! When he died in 2009, Peek, who loved to read at his local library, reportedly knew more than 12,000 books by heart.

Peek's brain had another unusual feature. Normally the cortex of the right and left hemispheres are connected by the *corpus callosum* – a band of millions of axons deep inside the brain, which allows neurons from the two hemispheres to communicate. Peek was born with no corpus callosum, so how his brain was able to coordinate information from the two hemispheres remains a mystery.

Einstein, on the other hand, had a corpus callosum with more connections than most people have, but we do not know whether this contributed to his brilliance. Although Peek's IQ was 87 compared to Einstein's 160 or so, brain scientists have yet to determine whether smart, creative people generally have a larger-than-average corpus callosum.

A PEEK INSIDE THE SKULL

Fortunately, doctors and brain researchers like me no longer have to wait for our patients – or geniuses like Einstein – to pass away before we can study their brains. If we did, we would know little more today than we did 50 years ago. Thanks to emerging scanning techniques, Peek could let scientists peek inside his skull and was aware that his own corpus callosum and parts of his cerebellum were absent.

In 1971 a group of engineers made the first major breakthrough in brain research when they invented the CT or CAT scan. 'Computed tomography' imaging can compile numerous X-rays of the skull, taken from many different angles, into cross-sectional images of the brain's interior. This technique was a revolution, for doctors and for patients. Suddenly tumours, haemorrhages and abscesses could be seen *without* opening the patient's skull.

The 1970s brought another revolution: MRI scans. Instead of X-rays, 'magnetic resonance imaging' exploits the fact that hydrogen atoms – present in virtually every molecule in our body – are ever so slightly magnetic. Think of the body's hydrogen atoms as billions of tiny compass needles. As you sit reading this book your body is not magnetic, because all the tiny compass needles are pointing in random directions, cancelling out each other's magnetic fields. But if we put you in the strong magnetic field of an MRI scanner, many of these compass needles would start to point in the same direction, making you into a sort of magnet with a south pole and a north pole.

Inside the MRI scanner, a radio transmitter nudges the countless compass needles in your brain simultaneously, measuring the radio waves emitted as they swing back to their pre-nudge position. The radio signals from the molecules in your brain reveal the chemical environment of their hydrogen atoms. This means MRI scans show fine-grained details in brain tissue that CT images cannot show – details a doctor might overlook unless a deceased patient's brain were removed and dyed slices of it examined under a microscope.

EINSTEIN'S BROWNIAN MOTION

The year 1905 was a memorable one for Albert Einstein: He published his theory of special relativity – the famous equation $E=mc^2$ – and his theory of the photoelectric effect. Less famously, he also formulated a mathematical description of 'Brownian motion', the tiny, dancing movements observable, for instance, in dust particles in a ray of light, caused by their constant collisions with much tinier molecules in the air.

Little did he know that decades after his death his discovery would revolutionise scientific studies of the intricate three-dimensional network of axons that jointly form the brain's white matter. Imagine a scientist in Einstein's day trying to follow, say, a bundle of axons from the corpus callosum to each end point in the cerebral cortex. This would require days or weeks of painstaking dissection, much like excavating a garden to follow the path of a thin, fragile tree-root among thousands of other roots in the ground.

In the 1990s scientists realised that axons, with their fatty myelin sheaths, affect the Brownian motion of water molecules in the brain's white matter, thus letting us determine the direction of the axons based on the water molecules' preferred direction of movement. Today, using computer software and MRI, which is sensitive to Brownian motion in water molecules, we can reconstruct the conductive pathways in the white matter in a living brain. Neurosurgeons now employ this technique to plan brain operations, as it helps them avoid severing vital nerve bundles during surgery.

In 2007 a Danish physicist named Sune Jespersen mathematically described how the infinitesimally thin dendrites and axons of brain neurons affect Brownian motion in water molecules in the cerebral cortex. Using his work, we can now quantify the countless extensions Ramón y Cajal observed in the brain's grey matter a century ago. And instead of using microscopes and slices of dead brain, we can use MRI to study microscopic changes in the cerebral cortex in real time – as our brains develop and learn new skills.

THE SUBCONSCIOUS BRAIN

WHILE YOU SLEEP

Now let us dive deep into your cranium. At the bottom of your skull is a hole that would accommodate a walnut. This is where your spinal cord, protected by the bony spine, reaches up into your brain. Here, the nerve fibres in your spinal cord blend into your brainstem, which stretches up to a point about midway between your outer ear canals before branching out into your cerebral cortex, rather like the stem of a cauliflower.

The brainstem is a superhighway for information going to and from the brain. It mainly consists of bundled nerve cords that pass through tiny holes in the skull or via the spinal cord to connect your cerebrum with the muscles in your body and face, and with the pain and tactile receptors in your skin. Information from your inner ears about sounds and balance also reach your cerebrum through the brainstem.

Your brainstem is constantly at work – even while you sleep. In fact, your brainstem decides how awake and how conscious you are, and keeps you breathing while you sleep. It also adapts your pulse, blood pressure and respiration to suit your body's oxygen needs, and it activates digestion while you are at rest. Your 'conscious brain' does not need to be aware of all these complex bodily functions, so your brainstem helps run things without inconveniencing your cerebrum.

IN FOR A PENNY, IN FOR A POUND?

The British psychologist Chris Frith and his French colleague Mathias Pessiglione designed an elegant experiment in 2006 to investigate how areas deep inside our brains prioritise our actions – unbeknownst to us. They asked test subjects to squeeze hard on a handle, measuring the pressure the subjects applied in the squeeze. Just before this task the subjects were briefly shown the reward for their trouble: a British pound, or just a penny – roughly two dollars, or two cents.

If we see a picture or a word for just a few hundredths of a second, our cerebral cortex does not register what we saw. Even if people are *told* they were shown something so briefly, they cannot point out the specific picture or word when they see it in a line-up. But the deep regions of our brain are much faster, and able to perceive words or images without us being aware of it.

The experiment showed that subjects squeezed the handle harder when promised a pound rather than a penny, even when the reward was shown in such a brief flash that they could not know its size. In other words, our subconscious brains give tasks higher priority if the potential reward is larger – whether we actually *know* it is larger or not.

Obviously, 'reward' is a very broad concept. Findings suggest that our brains basically perceive sex, food, touch and stimuli, including music, as rewards and have learned over time to perceive money and status as types of rewards, too.

In brain research we prefer to use monetary rewards in our experiments. Cash is useful to everyone and its value is easily determined, whereas offering test subjects a diverse array of 'brain-friendly' rewards would make our work infinitely more complicated.

TURN UP THE JUICE, PLEASE!

The parts of the brain that made Frith and Pessiglione's test subjects put more effort into their task when the reward was larger are part of our *limbic system*. There is a massive influx of information reaching the brain through our brainstem, and the limbic system serves as our front office, switchboard, emergency call centre, case-worker pool and secretariat. One part of this system, the *hippocampus*, reviews all incoming signals to decide whether the new input is so important that it merits being archived in our memory.

The limbic system's ability to manage behaviour was discovered a good 65 years ago by the American psychologist James Olds and the British engineer and brain researcher Peter Milner. Serendipitously, they noticed how their laboratory animals oddly kept prioritising seemingly pointless activities.

Olds and Milner had embedded electrodes in their

rats in certain parts of the *nucleus accumbens*, a part of the limbic system deeply embedded in the front of the brain. Electrical currents can stimulate nerve-cell activity, and Olds and Milner were hoping these electrodes might help them identify the parts of the limbic system that make up the brain's alarm centre, frightening the animals and making them flee.

The scientists activated the electrodes when the rats were in a certain corner of their cage, expecting the animals to scurry away. To their surprise, they found the exact opposite: The rats preferred to stay in the corner where their brains were being stimulated.

Olds and Milner proceeded to investigate whether their lab rats were willing to work to get current to flow through their brains. They equipped the rat cage with a switch the rats could use to activate the brain current themselves. The results were astonishing. The rats tapped the switch again and again, and some were so insistent they collapsed from exhaustion. The scientists even discovered their rats were willing to give up food and drink just to obtain little jolts of 'electrical juice'.

Evidently, the limbic system can make rats prioritise any task *above* their own survival. Quite a chilling thought. Tell your hippocampus to store it – this image of a rat willing to tap a switch until it dies of starvation. You will need it later to understand why dependency on, say, gambling or drugs or food can cause distressing changes in a person's behaviour.

Today, we know the electrodes in the Olds and Milner

experiments made neurons in the rats' nucleus accumbens release large amounts of the neurotransmitter *dopamine* each time a rat activated the switch. When an animal or human is faced with a task, the level of dopamine released by their limbic system decides the level of motivation, energy and persistency they feel. The electrode impulses made the rats' brain cells release so much dopamine that they decided to dedicate all resources to activating an otherwise trivial switch, even though doing this meant the brain's owner would die of hunger or thirst.

Humans, too, willingly invest time and energy in a certain behaviour if we see a chance of reward in the short or long term. Otherwise why would so many of us spend years in tiny dormitory rooms, eating cheap spaghetti dinners and foregoing new clothes and other creature comforts (available only at weekends back home), were it not for the academic degrees awaiting us at the end of all this hardship? But endure it we did, and do, because we believe education is the key to a brighter future.

The limbic system uses our experience to predict what reward our actions will fetch. Our brain then compares the anticipated reward with the actual reward. If an action results in a larger reward than anticipated, the limbic system releases more dopamine, motivating us to repeat the action. That is why we call dopamine release a 'learning signal': It makes us repeat and learn from certain types of behaviour. Now join me, if you will, in a small exercise and feel for yourself the brain's reward-system computations and dopamine release.

Imagine we have bought a lottery ticket. In a few seconds the winning numbers will be drawn, live, on television. First, notice that your brain does not merely consider factual information when computing the probability of winning millions. Obviously, lottery operators always pay out less prize money than they make selling lottery tickets, so the only reason we are sitting here with our lottery ticket is that adverts or stories about huge wins have given us an unrealistic expectation – mathematically speaking – of our own chances of winning. This expectation, paired with the fact that by purchasing a lottery ticket we actually have an itty-bitty, teeny-tiny chance of winning millions, may have made us decline a nice visit to our in-laws. Instead, here we sit, tense with anticipation, in front of the TV screen.

As the lottery-draw host reads out the numbers, one by one, we can physically sense how our lottery motivation depends on our dopamine level, which in turn depends on the difference between the anticipated reward and the actual reward. Feel the slight decline in excitement when we cannot find the first number on our lottery ticket. Our hopes are still high, though ... until at last we are forced to admit that our investment was wasted. Now, suddenly, the reward is lower than the brain anticipated, causing dopamine levels to fall below normal.

When our dopamine level reaches such a low point, we feel frustrated and lose our motivation. The conscious brain functioning in our frontal lobes may make us solemnly promise ourselves never to buy a lottery ticket again, from this day forward. The lottery companies know this, of course, so they make sure they regularly pay out titillating little rewards that make us feel like we *almost* won the big jackpot. This feeling releases large amounts of dopamine, and lo and behold! A week later, that jolt of juice to the brain will once again prevail over our conscious decision never, ever, to play the daily numbers again.

"AND FINALLY, MONSIEUR, A WAFER-THIN MINT"

Have you ever seen Precious, the ill-starred but indomitable heroine in the 2009 film of that name? Or the gluttonous Mr. Creosote in Monty Python's 1983 film *The Meaning of Life*, who ends up accepting a "wafer-thin mint" from the smarmy head waiter? Both these examples, although extreme, illustrate a general point about cravings and unhealthy eating. We all know the feeling. We don't really *want* to eat. We don't *need* to eat. Somehow we just feel *compelled* to eat. Yielding to this urge brings us no pleasure or enjoyment. We simply 'feel like' eating or grabbing a snack, and so we do.

For many years, brain researchers believed dopamine was a 'feel-good' neurotransmitter, and the parts of the limbic system that release dopamine are often referred to as 'the reward system'. However, the British–American brain scientist Kent Burridge has shown that mice and rats can enjoy sweet foods even though their neurons cannot release dopamine. Instead, dopamine seems to create the urge that motivates compulsive eaters and habitual snackers, the same urge that made Olds and Milner's rats keep tapping their brain-current switch. They simply could not stop themselves. And remember, eating did not make Precious or Mr. Creosote good-humoured or satisfied.

This has made many scientists and doctors revise their views on why it is so hard to overcome dependency and addiction. Gambling, alcohol and drugs like amphetamine, cocaine, heroin and cannabis can apparently affect and dominate the release of neurotransmitters in the human limbic system, taking control of our subconscious motivational system and perpetuating our behaviour – even though we no longer enjoy it and our conscious brain knows we may lose our job, family, livelihood or health as a result.

THE JOYS OF SURVIVAL

In 2010, psychologists from Harvard recruited over 5,000 test subjects whose mobile phones would be outfitted with an app. The researchers wanted to know what people were doing, and how they were feeling at the time, using an ascending scale of 0–100, from "very bad" to "very good". The app posed these questions at random times of day, and the study cohort consisted of female and male subjects around the world, aged 18–88, from all walks of life.

The results showed that certain activities were clearly linked with the subjects' happiness score. Making love scored highest, an unrivalled first. No surprise there.

The runners-up in feel-good activities were: exercise, playing, speaking with friends and family, listening to music, walking and eating. The scientific conclusion? People are most content when 'living', or more correctly 'surviving': procreating, strengthening social ties with their clan or group, eating and fortifying their body.

In recent years the Danish brain scientist Morten Kringelbach has been working with Berridge to study what happens in the brain when we feel happy and enjoy our life. They found that the neurotransmitters called *cannabinoids* and *opioids* produce a feeling of joy in our brains. Dopamine, on the other hand, produces urges and cravings and, in conjunction with other neurotransmitters, makes sure we learn the behaviours that initially triggered our elation.

As you may have deduced from their chemical names, the brain's own feel-good neurotransmitters are related to the contents of cannabis and opium. Opioids have recently come into focus as the root cause of pain-killer addiction, a problem that has reached epidemic proportions in certain areas. Now you know part of the explanation: Besides relieving pain, opioid drugs can generate a sense of joy and even euphoria. The brain will therefore tend to perpetuate opioid drug use, even after the user's pain subsides.

MAPPING THE CEREBRUM

POOR TAN

On 26 July 1830 the citizens of Paris took to the streets, fed up with their ruler gagging the press and dissolving parliament. Their call to arms heralded the July Revolution, where barricaded streets and clashes with armed soldiers and gendarmes lasted just 'three glorious days' yet overthrew France's unpopular monarch, King Charles X.

We do not know whether our next protagonist, Louis Victor Leborgne, actively fought in the streets, but we do know that he lived near Paris City Hall, the site of several violent clashes on 28 July. A powerful blow to the head may explain why Leborgne, then a young man, began to suffer from epileptic seizures.

Shoemaking was a growing cottage industry in Paris, and Leborgne, who turned 21 the week before the July Revolution, made a living carving wooden lasts for local shoemakers. Despite the seizures he was able to ply his craft until, aged 30, Leborgne suddenly, sensationally, became part of medical history when he lost the ability to speak. He did remain able to say one syllable: 'tan'. This he repeated so often that it became his nickname.

Perhaps an epileptic seizure caused a small cerebral

haemorrhage that destroyed part of his brain? At any rate, his unexpected handicap was a personal catastrophe. His inability to communicate made work impossible. Having neither wife nor relatives to care for him, after several months he was admitted to Bicêtre, a hospital on the outskirts of Paris.

Ten years later his health began to deteriorate, and the right side of his body gradually became paralysed. In 1861, poor bedridden Tan's paralysed side became gangrenous. It was after his transfer to the hospital's surgical ward that Leborgne met the doctor who would make his brain world-famous: Pierre Paul Broca.

BROCA'S DISCOVERY

Two years earlier, Broca had founded the Anthropological Society of Paris, a forum where the city's learned men could discuss important scientific theories. Broca was fascinated by Darwin's theory of evolution and the 'question of localisation' – whether certain parts of the brain handle certain functions, or whether speech and other functions were seated across the entire brain.

This idea of brain function localisation was an offshoot of *phrenology*, a popular scientific discipline in the 1800s that assumed a person's skull could reveal how well-developed their brain was. The basic hypothesis – which was utterly flawed – was that scientists could 'read' a person's moral standing and abilities based on the shape of their cranium.

When Broca heard of Leborgne's unusual handicap, his medical mind instantly saw its huge potential. At long last, the question of localisation could be put to the test, in practice! If the seat of speech lay in a certain spot, then logically Leborgne's brain must have been damaged in that spot. While examining his afflicted patient, Broca understood from his gestures and 'tan-tan' utterances that Leborgne understood everything he said and seemed intellectually intact.

Leborgne died soon after, and Broca was able to examine his brain in detail. Upon removing the crown of his skull, Broca made his great discovery. Beneath Tan's cranium, at his left temple, the cerebral cortex had sustained damage, leaving a cavity the size of a small walnut on the brain surface, just as the theory of localised brain functions had predicted. Broca later examined several more patients and found that their lacking ability to form words almost always resulted from a cerebral-cortex injury to the frontal lobe, near the left temple.

We now know that a small percentage of people, most of them left-handed, have their language function seated in this area on the brain's *right* side, or both sides. This is important for stroke victims or brain-surgery patients, since damage here can severely impair their ability to speak. Like Leborgne, they will have a medical condition we now call *expressive aphasia*. And the piece of cerebral cortex Leborgne was missing? It was named *Broca's area* after the doctor who revealed its function.

The German physician Karl Wernicke was inspired by Broca's studies. In 1873 he examined a male patient who could hear and speak but not understand spoken or written words. This was the exact opposite of Leborgne, who understood what people said but could not speak. After his patient died, Wernicke examined the man's brain and found that his comprehension problems were due to damage not in Broca's area, but in an area roughly a hand's width further back, where the temporal lobe meets the parietal lobe – now called *Wernicke's area*.

DOES THIS TICKLE?

In the early 1900s brain surgery grew increasingly common. Surgeons had learned how to avoid the infections that nearly cost Phineas Gage his life, but they faced a new problem, namely avoiding damage to important areas of the brain while trying to remove diseased tissue. Unlucky patients ran the risk of waking up with paralysed limbs or compromised language faculties.

Then, in the early 1930s, a bold Canadian neurosurgeon named Wilder Penfield invented a radical technique. During surgery he would stimulate the exposed cerebral cortex with a weak electric current – while asking the patient what sensations this caused. Yes, the patient would be awake. This may sound horrifying – being awake while a surgeon fiddles with your brain. However, the brain itself cannot sense pressure or pain, hot or cold, so the biggest hurdle is getting used to the idea of it.

Penfield's technique allowed him to navigate his scalpel by patient reactions. If the patient felt a tingle or jerked an arm or leg, Penfield knew he had to be careful. In this way he could reach the sick brain tissue without leaving patients paralysed.

During operations, Penfield also mapped his patients' motor and sensory strips in great detail. He discovered that nerve fibres to and from the tongue and throat are located at the very bottom of the motor and sensory strip, just above the temporal lobe. A bit further up he located the cortex that controls the mouth and lips. Neurons in front of the central fissure control our facial muscles when we make faces and purse our lips, and the nerve fibres that transmit sensory input from the lips and mouth end just behind the central fissure – so actually, Penfield mapped out how the brain helps us enjoy a nice, long kiss.

BERTINO'S POOR NOGGIN

In late July 1877, a 37-year-old Italian peasant named Michele Bertino from the village of Varicella, near Turin, inadvertently became a celebrity in the history of cerebral cartography. As Bertino stood at the foot of the village bell tower ready to hoist a basket of bricks up to the workmen above, suddenly his world went black. A mason had lost his grip on a brick weighing three kilograms (almost seven pounds), which fell 14 metres (45 feet) and crushed Bertino's skull.

The stricken man was carried to the home of the village priest. The local surgeon, Doctor Ferrero, was present when he regained consciousness and found Bertino able to recall everything until the brick struck, and he described the patient as oddly unaffected by the severe trauma.

Ferrero cleansed the wound a few hours later, relieving the unfortunate peasant's scalp and underlying brain of seven skull fragments and remnants from Bertino's hat and the notorious brick.

Over the next few months the wound gradually healed. Bertino spent several weeks, late September to late October, at a hospital in Turin where news of the incident reached the physiologist Angelo Mosso.

Mosso was studying the pressure waves the heart creates when it pumps blood around the body. Most readers who have had their blood pressure taken will have heard of the two values measured: the high *diastolic* pressure that arises when the pressure wave from the heartbeat reaches your upper arm; and the lower *systolic* pressure while the heart rests between beats. Mosso used a special apparatus, a 'hydrosphygmograph', sensitive enough to measure how blood pressure varies as the pressure waves pass through the arm and equipped to record the values as curves on strips of paper.

Examining the almond-sized hole in Bertino's skull, Mosso noted that the cortex did not swell out through the aperture, as is normal during brain surgery. This meant Bertino's brain would not disturb the physiological experiments. Mosso was thrilled.

Bertino was not. He disliked all the medical attention, but he knew he would get headaches and grow ill from infection if he did not let them cleanse the wound each day. Mosso persuaded Bertino to visit the laboratory during his last week as an inpatient in Turin, resulting in the experiments that would make Mosso famous.

First, Mosso placed a pressure gauge above the hole in Bertino' scalp and a pressure gauge around his arm. He then asked Bertino to solve different tasks while the apparatuses recorded the pressure inside his skull and the blood-pressure values in his arm. "What is 8 times 22? What is 8 times 12?"

Mosso could see the pressure rise in Bertino's head as his brain did the calculations, and when he stated the result. The pressure in his arm, however, did not change. This had to mean that the blood vessels inside his skull were expanding as his brain worked. Mosso also noted an aberration recorded on the paper strip when nearby church bells chimed noon. He wondered: Was Bertino sad that he was unable to make the sign of the cross and recite his noontide Ave Maria, as was his habit? Mosso asked Bertino, who answered in the affirmative. Here, too, Mosso observed a pressure rise as Bertino wrestled with his moral dilemma, weighing his sense of religious duty against his promise to follow Mosso's instructions during the experiment.

NEURONS OPEN THE FLOODGATES

News of Mosso's fascinating results reached Britain, where the Cambridge physician Charles Smart Roy and his colleague Charles Scott Sherrington, a physiologist like Mosso, pursued the Italian's ideas. They outfitted a laboratory to study intracranial blood-vessel expansion in dogs, cats and rabbits, stimulating various nerves in the animals, changing the flow of oxygen and blood to their brains, and applying various chemical compounds.

In 1890, Roy and Sherrington published their remarkable results, concluding that the brain itself increases its blood flow to provide the extra nutrients needed during work. This principle – *neurovascular coupling* – would become the key to investigating brain functions over the next century.

Thanks to neurovascular coupling, today we can see which parts of the brain are at work. All we have to do is look for changes in its internal flow of blood.

In the 1960s, a Dano–Swedish research duo, Niels A. Lassen and David Ingvar, were the first to measure blood flow in various parts of the brain. First, they fixed Geiger counters in various positions on the skull of their test subject, then injected a radioactive trace substance into the brain's blood supply. The radiation readable on the Geiger counters enabled them to measure how much blood was flowing through the brain just beneath the skull.

In October 1978 their research adorned the cover of the prominent journal *Scientific American*, inspiring a whole generation of researchers who would map the brain in detail over the next few decades. The image sketched out the left brain hemisphere of a person reading aloud, lit in shades of warm orange and red at Wernicke's area, Broca's area and the neurons in the motor strip that control the speech muscles – just as Penfield's operations had shown. A century after Bertino's awful mishap, neurovascular coupling finally enabled scientists to map out the brain functions going on under our thick skullbone.

HANG ON: I KNOW YOU!

Humans are incredible at facial recognition. We can even identify most of our childhood classmates at a reunion twenty or forty years later.

The Lebanese neuropsychologist Justine Saade-Sergent found out why. The visual cortex is linked to a dedicated area that helps us recognise faces. This is just one of many special areas identified since the mid-1980s. Engineers had developed PET technologies with detectors so sensitive to radioactivity and electronics so advanced that doctors could capture blood-flow changes deep inside the brain and map its functions in ever-greater detail.

Positron emission tomography scans also let researchers measure the release of neurotransmitters, including dopamine, in various parts of the brain. When experiments showed that the brains of young computer users release huge amounts of dopamine during gaming, it caused quite a stir. But this is no revelation to any parent who has ever asked a child to leave an electronic game to come and eat. Making mental contact is all but impossible, and the parallel to the Olds and Milner rat experiments is obvious in the willingness to forego food and drink to press the buttons that keep their dopamine pumping.

MAGNETIC BLOOD – AND A REVOLUTION

In the late 1980s the Japanese physicist Seiji Ogawa

discovered that MR images of animal brains turn out slightly darker if blood oxygen content is low. He dubbed this 'the Blood Oxygen Level Dependent (BOLD) effect' and proposed using MR scanners to monitor brain activity, as brain cells use more oxygen when they are at work.

Around this time the American physicist Ken Kwong was working with the BOLD effect in humans. In mid-May 1991 he succeeded in measuring brain work using an MR scanner but no trace substances at all. Thanks to neurovascular coupling, he became the first person to see the visual cortex light up on MR images as he showed flashes of light to a test subject in the scanner.

At the annual MR congress in San Francisco that year, Kwong's boss, Tom Brady, cut into his own keynote speech to show us all Kwong's amazing findings. A collective gasp of excitement went through the audience, and my guess is that all those present that day, including yours truly, still vividly recall the moment when this spark ignited a revolution in our efforts to map the brain.

As I write these words, brain scientists around the globe are using MR scanners to map the network of neurons which, working together, create our thoughts. One outcome is that many believe that when recalling a real, experienced event, our brain works differently from how it works when fabricating a memory. Some researchers envision a new type of lie detector for police investigations, as this new method is currently thought to have a hit rate of about 95%.

THE LEARNING BRAIN

SYNAPTIC RECALL

The brain acts as a sort of hard disk that stores experience, knowledge and new skills. The hippocampus plays a vital role in committing things to memory. Patients with a damaged hippocampus often recall their childhood and events up to the time of injury, but very little after the injury.

In the late 1960s the Norwegian muscle physiologist Terje Lømo and the British brain researcher Tim Bliss discovered the mechanism that enables the brain to remember, a phenomenon called *long-term potentiation*, LTP for short. While using brief pulses of electricity to stimulate neurons in the cerebral cortex that overlies the hippocampus, Lømo looked at the signals these pulses caused the neurons *inside* the hippocampus to send. He found, to his surprise, that if the neurons recognised the pulse from previous experience, they emitted a stronger signal next time they received an electric pulse, **even several** hours later.

Meanwhile, Bliss was searching for signs that neurons forge stronger connections if they repeatedly send certain signals to each other. This would explain how networks of neurons gradually learn, say, a movement we repeat many times. Thanks to their in-depth studies of LTP, together Lømo and Bliss found out that the synapses enable brain cells to form temporary memory circuits.

LTP is merely the first step in an astoundingly complex process. Over time, synapses form permanent networks among brain cells that remember things we experience and learn – as presaged by Ramón y Cajal. Picture walking in a green meadow with grazing cows, singing skylarks and fragrant flowers. These impressions are briefly retained as impressions of sight, sound and smell in various parts of the brains by our short-term memory. Over the ensuing hours or days the hippocampus can 'aggregate' them and re-create the experience in our thoughts. Then, if a memory path is important, LTP, along with chemical messengers and special proteins, can help our brain store it – forever.

This phenomenon is also important to small children learning to coordinate their movements. Imagine an infant in a crib, eyes firmly fixed on a rattle dangling above. Waving its arms, the baby tries to grab the rattle. In theory the infant's brain can already do this, as the brain cells that process visual input are linked to the areas that plan and perform arm and hand movements.

But the baby has only just begun to get the right neurons to work in the right order. Now imagine how each time its flailing arms move closer to the rattle, the infant gets a tiny reward from its limbic system. This teaches it to repeat the movement, while LTP reinforces the nerve connections each time the infant repeats the movement. The baby's hard work gradually selects the neuronal network circuits that solve the task at hand, and its movements become more intentional, coordinated and precise.

COPYCATS ALL

We often learn by imitating others. Just think of that new routine at the gym, where you follow your instructor's movements. Our brain helps us do this. Say you watch someone pick up a glass and drink from it. The areas your own brain uses to do likewise will change their electrical activity as you watch, programming it to imitate the actions you see.

Fortunately, your brain and spinal cord have a brake mechanism, so you need not keep pace with everyone else drinking around you. Even so, imitation is a powerful force. Adults spoon-feeding babies often impulsively open their own mouths, and even infants can mimic the mouth movements they see in others. Perhaps spoon-wielding adults unwittingly use this reflex in children to ease the process?

The brain's ability to copy movement gives it a huge advantage when learning new skills. A baby who sees its parents' open-mouthed faces hardly knows it has a face with a mouth of its own that it must open to eat and feel sated. But the baby's copycat of a brain immediately activates the right neural circuits, and thanks to LTP the baby quickly learns to open its own mouth when the spoon draws near.

You may be wondering how a baby's brain can translate the mere sight of the open-mouthed parents into the action of opening its own mouth. This also puzzled the Italian neurophysiologist Giacomo Rizzolatti in the 1980s. In a series of tests, he found that monkeys' brains have certain neurons that emit identical electrical signals whether the monkey is just watching another monkey eat a banana or actually eating one itself. Later, other scientists found such *mirror neurons* in humans, too, which may be what train infants' brains to learn by imitating grown-ups.

Some brain scientists believe mirror neurons also help us to decode intentions and emotions in others **and know,** almost instinctively upon seeing a face, whether a person is happy, sad or angry. Thus far, brain scans show that certain regions of the human brain exhibit the same activity whether we are looking at a picture of a face expressing a certain emotion or making that expression ourselves.

THE PAYOFF OF PREDICTING

An infant learns from experience to master the arts of gripping, crawling, walking and so on. When it comes to sensory input, a baby's brain works equally hard to process the impressions it gathers as it learns to navigate in the big wide world.

Take the sense of sight. As we explore the world, the retinas send lots of impulses via the optic nerves to the occipital lobes. There, one might think the visual cortex uses its full computing power to process all visual impressions as they reach it, so that we see everything the retina registers.

Apparently, however, the visual cortex is much more advanced. It helps the brain maintain an internal model of our surroundings, and constantly checks for obstacles and dangers as we navigate and move around. To do this, the visual cortex seems to combine its internal model with our experience to predict which visual impressions it will receive next from the optic nerves. Should its predictions fail to tally with reality, the brain can update our internal model to be more accurate and detailed – and alert us if unexpected sensory inputs arrive.

It may seem odd that the visual cortex makes such complex computations instead of simply taking in what is happening. To understand why the brain must be proactive, imagine standing in a large city square. There are no cars. Shops and cafés surround the perimeter, and people slowly flow through in various directions. Some hurry while others stroll along, stopping now and then to contemplate the buzzing spectacle of life.

Three children are playing with a ball, which occasionally bounces in among the crowd. Now imagine you decide to cross the square to a café. If you had an 'observing brain' you would have to constantly, consciously monitor the flagstones in front of you for pigeons, balls, benches and other pedestrians. If something got in your way, you would instantly have to change direction while still aiming for the café, your target destination on the other side of the square.

But humans have a 'predicting brain' that can do this job for us. On your many walks together, you and your brain have already seen thousands of pedestrians moving and balls bouncing on hard surfaces, so in a fraction of a second your brain can estimate their direction and speed. That is why your brain does not constantly have to keep an eye on everything. Instead, it can occasionally check that the people and balls are roughly where it had predicted they would be, and it can even plan a route to avoid the ball-bouncing children.

The payoff is this: Your brain and eyes, combined, can act as a sophisticated autopilot, monitoring and adjusting your course across the square. This leaves you free to gaze at the lovely façades, the portrait-painter by the fountain, and the flower containers in full bloom, imprinting them as detailed images in your inner model of the square. If nothing unforeseen happens, you will reach the café as planned – having let your automated, predicting brain take care of the journey while you took in the sights. Brain researchers call this mechanism *predictive coding*, and it enables the brain to navigate based on experience and just a few visual impressions – while you actively use your senses for more exciting things.

YOUR INNER FILM REEL – AND ITS WHITE SPOTS

Meanwhile, as you cross the square you suddenly catch

sight of a street magician with a circle of onlookers. He deftly makes balls and coins vanish before the eyes of his incredulous audience. You stop to watch. Although you focus intently on his hands and props, you cannot work out the secret of his sleight of hand. How does he do it?

Well, there are strengths *and* weaknesses of your brain's ability to guess what you are seeing. Scientists believe the outermost cell layers of the visual cortex predict what the world looks like and how it will behave, while its deeper layers compare these predictions with the visual input constantly flowing into the cortex. If the discrepancies between prediction and real world become too large, the deep neurons send a message about the deviations to the outer neurons, which then proceed to improve their model. In this way the brain creates an inner reel of film that represents our surroundings but is constantly being updated with details that make it ever more precise.

Back in the square, you experience the downsides of predictive coding as the street magician deceives your senses time and time again. The brain inevitably tries to update its image of the street magician in areas where visual inputs are rapidly changing. The street magician knows this, so he makes sure to draw your brain's attention away from the places where his tricks are working their magic. That is why your brain fails to update your inner image in the area where the magician's left hand smoothly hides one of his props – because it is busy updating the area where his quick-moving right hand fans out a deck of playing cards. Simply put, the trick never becomes part of the film sequence your brain is creating of the street magician, no matter how firmly you focus or stare.

Resuming your walk across the city square, your sightseeing is once again interrupted as you register, out of the corner of your eye, several people stopping in their tracks. Next thing you know, you are observing a small procession of people in colourful medieval costumes entering the square from a side street. What just happened in your brain is really quite remarkable.

While you were consciously preoccupied with the lovely façades around the square, your predicting brain discovered that something was amiss. Several other pedestrians within view had not reached the location your brain had predicted. Instead, they had come to a standstill. Also, in the mass of ebbing, flowing humanity, suddenly there were clothing items and colours that seemed incongruous on a summer's day in the early twenty-first century.

Back in 1978, the Finnish psychologist Risto Näätänen discovered that the cerebral cortex emits a special *prediction error signal* if our surroundings suddenly change. He studied how the auditory cortex reacts when analysing a series of sounds that subtly changes in tone frequency or rhythm as it plays.

If you have ever seen – or rather, heard – the initial auditions for a show like *Britain's Got Talent*, you will know that humans immediately notice if the voice singing 'Livin' on a Prayer' is not at least a rough approximation of Bon Jovi's original rock hit. Some people even physically cringe at disharmony, so it is hardly surprising that scientists can measure such discomfort in our brains. What *is* surprising about Näätänen's error signal is that our auditory cortex subconsciously registers deviations in the soundscape, even when our attention is focused on something completely different.

In other words, the auditory and visual cortex areas constantly assess inconsistencies between the incoming impulses and their own predictions. Näätänen's error signal measures the difference, and if it is powerful enough it commands your attention. Your vigilant, subconscious brain is attuned to your surroundings and warns you if something sufficiently unexpected happens. In this case, a mere pleasant surprise: people appearing in a city square in clothing that is anachronistic by several centuries.

CAN YOU RAED TIHS?

The brilliant German physicist and medical doctor Hermann von Helmholtz proposed as early as the 1800s that our subconscious mind constructs an inner model of the world to explain the sensory input we receive. The British psychiatrist Karl Friston has pointed out that the brain may even control our movements by means of the sensory signals our muscles and tendons send back to the brain. In short, the brain activates movements and controls them by checking that they feel right.

Your brain's ability to subconsciously predict events in the world around you – only alerting your conscious brain when it perceives anomalies in the sensory input – gives you huge advantages. It enables your subconscious brain to take over more and more tasks as you learn new functions. This 'automation' frees up computing resources, allowing your conscious brain to learn even more new things, which it then automates, and so on and so forth.

The brain is constantly storing information that improves its predictive abilities. When learning to read in school, first we learn letters, then how letters form words, then how to spell words correctly. But for our brain, there is more to solving the task, and it also follows rules that are different from the ones you learned in school. In spoken language, children use fun systems like Pig Latin and Ubbi Dubbi. And think of how easily learn to make out accents and dialects you can, or understand Master Yoda of *Star Wars* fame. Put this now to the test we shall!

The next paragraph may seem illegible at first, but give it a go. You may just be able to decipher it.

Can you raed tihs? Heer I hvae julmbed smoe of the ltertes in the wodrs. The frsit and lsat lretets are mlotsy in the rhigt plecas, but tohse in bweeten aer in rodnam oredr. Tsih teaks soem gtnetig uesd to, but wtih a bit of pirctace, yuor biarn fnids a sytesm. It divoscres taht the txet hlods mineang, as lnog as you dn'ot epxcet the mdilde letrets of ecah wrod to ocucr in teh uasul odrer. Atefr a wlihe msot popele are albe to raed txet 'ecpyrtned' in tihs way wtih olny mosedt dfitfilucy. Yuor barin can rignecose a hgue nmebur of wdros, and it deos sith so wlel it can wrok out waht tehy are sposuped to sya, eevn tohguh the slenpilg is taltoly wnorg. Celevr, eh? Srnotg in yuo the focre is!

BE
PREPARED

A BRAIN IN DIRE STRAITS

'Be prepared!' Famously applied to scouts, this motto is equally fitting for brains, as one of their crucial functions is to gather information and develop skills that can keep us out of danger – or help us deal with it.

For our ancient ancestors, being prepared to handle attacks from predators was often a matter of life and death. When suddenly in dire straits, they had to act quickly and appropriately. How exciting it would be to study coping mechanisms in the brains of people in mortal danger. However, for obvious reasons no scientist would ever want to plan or be allowed to conduct such an experiment. Instead we can use television and the Internet to see how people react when they *think* they are in mortal danger.

A quick online search for 'scare pranks' calls up a torrent of videos with pranksters frightening the living daylights out of friends and relatives. Studying the outcomes is fascinating. Some victims simply freeze. The rest either instinctively attack the prankster with startling force or take flight with amazing speed. These videos clearly show how the victims' fast, impulsive reactions and loud screams almost always fluster the attacker.

When in mortal peril, we have no control over the

reflexes that make us freeze, flee or fight for our lives. Our *amygdala* – an almond-shaped brain area tucked deep within the temporal lobe – makes these decisions for us. The amygdala is what kicks in when we think we see an axe-wielding marauder lurking behind the kitchen door on a dark and stormy night, or when we feel intense fear or rage.

The amygdala's role as our spring-loaded, knee-jerk lifesaver when danger threatens has given it a special privilege. If the visual or auditory input from our surroundings is dangerous enough, the amygdala can instantly take command and make us stand stock-still, punch and kick, or run away at top speed. This prevents any delays our conscious brain would cause while trying to come up with a plan to save our skin.

Meanwhile, the amygdala causes our pulse, blood pressure and muscle-tissue circulation to increase, preparing us for the ordeal ahead, initiating actions as we sense fear but before we know the cause.

What is more, our amygdala not only reacts instinctively to dangerous situations. It remembers them, too. So if we see, hear or touch something later that reminds us of the first time we came across a specific threat, the amygdala can go straight to red alert, improving our chances of coping, should the same situation arise again.

The amygdala's red-alert function is not always appropriate, or desirable. Sudden loud noises or flashing lights can activate the amygdala, for instance in a soldier previously involved in hostilities in a war zone. Not only will the amygdala call up inner images of explosions and injuries; it will also repeat the forceful mental and physical stress reactions the soldier had at the time, yet here the red-alert state – a post-traumatic stress reaction – does nothing to help the soldier.

I deliberately wrote "activate the amygdala" because the cerebral cortex is not necessarily involved here. This is because the amygdala can make us feel fear, anger or sadness without us knowing exactly what triggered the emotion. The cortex, however, cannot conversely control the amygdala, so the soldier cannot decide *not* to relive the stressful state over and over again. Scientists are working to devise a method that can delete bad memories – a daunting task, as they fear they might accidentally delete other memories too, or damage important skills.

Let us revisit your mental walk across the city square, just a split second before the error-signal event drew your attention to the people in medieval garb. You reflexively turned your head, reacting to an anomalous element in your field of peripheral vision.

Friston has suggested that our subconscious brain not only predicts the things our senses detect; it also ensures that it gathers better data for making its predictions. When you turned your head towards the unknown anomaly, your predictive brain made you do this, so it could gather better visual data for its work. If its data-collection mission had, instead, revealed a waiter moving café parasols among the pedestrians thronging the square, your subconscious brain would not have interrupted your sightseeing but let you pursue your trajectory to the café on the opposite side of the square.

So: The brain can create concord between its model of the world and the real world by gathering more information. The question is this: If there is discord, will the brain try to change the *real world*, or will it change the *model of the world* it has created?

Picture yourself taking a seat at a large dinner party. Each place is set with several glasses and multiple sets of cutlery. The middle knife by your plate is slightly misaligned, touching one of the other knives. Do you nudge the middle knife to straighten it – without thinking? Or do you not touch it until the main course is served? Many of us would likely nudge the knife to arrange the cutlery neatly, creating order and aligning things with our internal model of the world.

C4N Y0U RE49 TH15 T00?

Now let us see what happens if the brain cannot make its model fit. Imagine you are suddenly time-warped into a parallel universe. It looks very much like this universe, but with one obvious difference. The letters of the English alphabet look quite unlike the ones you are used to reading. The question now is: Will your brain stubbornly stick to its own model of the world – your normal alphabet, år my nårmøl Dænish alphøbæt – or will it construct a new inner world model and get used to the new rules? The answer lies in the next paragraph.

C4n y0u 4l50 r349 wh4t 1 h4v3 wr1tt3n h3r3? 1t 15 4 61t
h4r93r, 63c4u53 n0w 1 h4v3 r3pl4c39 50m3 0f th3 l3tt3r5
w1th 53v3n 91g1t5 63tw33n z3r0 4n9 n1n3! F0r 1n5t4nc3,
1'v3 5w4pp39 th3 f1r5t l3tt3r 1n 60th 0ur 4lphø6æt5 w1th
th3 num63r f0ur, 4n9 th3 f0urth l3tt3r w1th th3 num63r
n1n3. 1t 15 unl1k3ly th4t y0u h4v3 3v3r r349 w0r95 w1th
91g1t5 1n th3m 63f0r3, 50 1f, 935p1t3 y0ur l4ck 0f pr3v10u5
3xp3r13nc3, y0u 4r3 46l3 t0 r349 th15 t3xt, 1t 15 0nly
63c4u53 y0ur 6r41n h45 gr49u4lly 63c0m3 46l3 t0 939uc3
th4t th3 t3xt c4n c0nv3y m34n1ng 1f y0u 'r349' 50m3 0f th3
91g1t5 45 1f th3y w3r3 l3tt3r5. N0w y0u mu5t 49m1t, y0ur
6r41n r34lly 15 qu1t3 6r1lll14nt, 15n't 1t?!

This innocent letter–number game raises a more
serious question: Might the brain also disregard what
we know as 'right and wrong' if circumstances demand
it? Actually, Friston's theory predicts that a brain will do
everything it can to minimise discrepancies between its *own
model* of the world and the *real, actual* world.

Numerous psychology experiments confirm that we
humans are ready to change our norms and behaviour
significantly to avoid clashing with what our environment
expects. If a young person gets involved in crime, addiction
or extremism, we often blame it on 'bad parenting' or
'lack of moral fibre'. But if our own environment changed
radically, it is very unlikely that our brains would stick to
the rights and wrongs we learned as children. After all, we
have just seen it happily abandon our preconceived idea
of the alphabet to avoid leaving us utterly clueless in that
parallel universe we just visited.

LET YOUR MIND SOAR

FOCUS, PLEASE!

"Dad, what did I just say?" This question came from my daughter Bertha, then ten years old. She wanted to make sure I was listening to her news about the horses we knew from our frequent nature walks in the hills of Mols Bjerge national park in East Jutland.

Bertha's question exemplifies a sort of standard test. We ask such questions to confirm that our father, wife or friend is focusing on the here and now. If their reply is not satisfactory, they usually make excuses and explain that they were daydreaming, letting their mind wander, or thinking about something else.

Earlier, I mentioned a study in which Harvard psychologists used an app to ping their study participants at random times of day for information on what they were doing and how happy they were. They also asked subjects whether they were thinking about something other than what they were doing when they got the ping, which 47% replied they were. In other words, whether respondents were riding the bus, working in the office, listening to their favourite music or reading a novel, nearly half said their thoughts were elsewhere. Only one activity commanded undivided attention: Less than 10% of respondents who were making love when they got the ping were thinking of something else.

We do not decide which thoughts our brain initiates, or whether our inner film reel is accompanied by positive or negative emotions. The Harvard study generally found the subjects who were daydreaming when pinged to be less happy than those who were not. However, the contents of the daydreams by no means explained their mood.

We need not fear that daydreaming will ruin our mood. Daydreams probably help us run through scenarios and equip us to cope with them better later, if they arise in the real world. For me personally, daydreaming also seems to bring good ideas.

Daydreams are hugely challenging for scientists trying to map the human brain. Say we want to study what your brain is doing while you look at art. First, we put you in our MRI scanner. Then we show you a series of images where a 'blank screen' alternates with various modern sculptures. While we do this, the scanner creates BOLD images of your brain. Finally, we have a computer identify which areas 'lit up' as you were looking at the art. Thanks to the work of Mosso, Roy and Sherrington, today we are at liberty to assume that your visual exposure to art is what activated these brain areas.

The problem with using changes in scanner images to identify your art-processing centre is that your brain does not 'zone out' while you look at the blank screens between the sculpture photos. It daydreams, and the art may even affect the thoughts and associations arising after the sculpture photos disappear. To understand where and how art affects your brain, perhaps we really ought to be looking at the areas that light up *between* the sculpture photos?

Many brain scientists are now using MR scanners to study changes in brain activity *without* instructing the subject what to think about during scanning. The reason is that many of the areas that do *not* light up on brain-scan images while the person is solving a specific task are more active during daydreaming. Put differently, our thoughts and daydreams may arise in parts of the brain hitherto unstudied because we, the scientists, have mainly looked at the problem-solving brain doing tasks posed by us, not by the brain's host.

Several laboratories now use MEG, magnetoencephalography, to measure brain activity more directly. MEG scanners can measure the minute magnetic fields that arise when the neurons in the cerebral cortex send signals to other parts of the brain. Some day we may even be able to film the birth of a thought. In my book, that would be truly mind-boggling.

We may also seek help in unexpected places to study the daydreaming brain. For millennia, introspection and meditation have been important tools for philosophers and religious thinkers. Perhaps this is because, as noted above, the human brain spends much of its time letting thoughts fly high and free. This is interesting in the context of brain research because several meditation techniques focus on examining one's thoughts and sensing the world differently from how one's brain usually senses.

People often begin meditation by concentrating on bodily sensations, for instance how the air passes in and out, in and out, through the nostrils during breathing. You have already learned that our brain predicts such sensory impressions and only alerts us if they change. If a snowflake or a mote of dust tickles your nose, you may consider whether to suppress a sneeze. In other words, the predictive brain tells us about *changes* in sensory impressions rather than telling us about the impressions themselves.

During meditation, people also practise directing their attention to specific objects or thoughts. They may experience, for example, a cherry tree more intensely when they focus on observing its blossoms, leaves and branches in great detail.

You already know one possible explanation. The brain, by default, portrays a *generic* cherry tree on our inner film reel – and lets us enjoy the sight, even if we pass the tree in a fast-moving car. It takes time, however, for the visual cortex to 'upgrade' this image with the wealth of information and beautiful details that characterise a *unique* cherry tree in front of us. So, if we truly want to experience the world around us, we must be willing to spend time on taking it in.

ARISTOTLE GETS THE LAST WORD

It was back in 1637 that Descartes wrote: "I think, therefore

I am." This statement shows he had realised our senses can ultimately trick us. Theoretically, the physical world as we know it could be an illusion, so in his view the ability to think – and to doubt – was the only reliable basis for understanding the world. The reflections in this small book would likely have reinforced Descartes' scepticism of the world-images our brains create.

Fortunately, brain scientists, psychologists and philosophers are now working *together* to understand the human brain and consciousness. They hope to answer questions you may have asked yourself while reading this book: Am I my brain? Do I control my brain, or does it control me? And if it does, what about my own free will? Such knotty questions continue to baffle us. While there is a consensus that humans *feel like* they are controlling their own actions, when we look in detail at the brain's internal decision-making processes we find no decisive evidence that human beings do, in fact, possess a free will.

In the early 1980s, the American physiologist Benjamin Libet decided to investigate how the brain controls our actions. He wanted to study the time lapse between us *deciding* to act and *carrying out* the action we had decided to do.

Libet asked a number of test subjects to move a finger when they felt like moving it, having first applied two electrodes that would measure activity in their brain and finger muscles, respectively. Libet knew the brain electrode would register a *readiness potential* as early as half a second before the finger moved. The readiness potential may be caused by neurons planning the action that the test subject had decided to perform.

Libet used an oscilloscope with a rapidly circulating dot as a clock, asking his test subjects to remember where the dot was at the moment they made their decision to move. You are probably thinking, dear reader, that the series of events is predetermined: First the subject decides to move their finger; then the readiness potential arises; and finally the finger moves.

But when Libet analysed his results he reached an odd conclusion indeed: The cerebral cortex was already planning the finger movement *before* the test subjects reported they had decided to move their finger.

Researchers and philosophers are still discussing how to interpret Libet's findings and, not least, their consequences for our self-perception. Apparently the brain makes decisions for us, but it lets us think we are in charge.

While writing this book I had the pleasure of chairing a PhD defence at the University of Aarhus. The candidate was Mads Jensen, a promising young philosopher *and* cognitive neuroscientist whose dissertation built on experiments that yielded new, exciting findings on certain aspects of Libet's work. Two prominent international researchers – specialists in the field Mads was studying – were tasked with acting as opponents, asking him hard questions and making sure there were no holes in his work or his knowledge.

When they were convinced Mads was on top of things, one opponent asked the question I had been eagerly anticipating. "So, do we have a free will?" Mads very sensibly explained that some researchers still believe we do, adding that it is probably more a matter of us being free to choose what we do *not* want to do.

To sum up his point, our brains constantly work with possible scenarios and plans for our future, imminent and more distant. Gradually, as these plans mature, we can stop actions unsuited to the situation. Chances are you did not actively choose to read this book, but instead chose *not* to watch a film, empty your dishwasher or do something entirely different.

Aristotle lived in Greece almost 2,400 years ago, but to this day he is regarded as one of the most important biologists and philosophers the world has ever seen. He was wrong on one count though, claiming that our thoughts, senses and movements were controlled by our heart, whereas the task of our brain – by virtue of its considerable size and moist surface – was largely to cool the blood.

Aristotle did, however, point to an important distinction we too have reflected on. Despite the two thousand years between his time and mine, we agree that there is a huge difference between walking, seeing and hearing, and being *conscious* of doing so. Although he saw the brain as little more than an elaborate cooling system, I shall nevertheless let Aristotle have the last word with a quote from his *Nicomachean Ethics*, a treatise on the good life:

"To be conscious that we are perceiving or thinking is to be conscious that we exist."